3.97

D1737818

A Stepping-Stone Book

Why Things Work
A Book About Energy

by JEANNE BENDICK
Pictures by JEANNE BENDICK
with KAREN BENDICK

Parents' Magazine Press • New York

Text and illustrations copyright © 1972 by Jeanne Bendick
All rights reserved
Printed in the United States of America

Library of Congress Cataloging in Publication Data

Bendick, Jeanne.
    Why things work.
    (A Stepping-stone book)
    SUMMARY: Discusses the sources and uses of different types of energy.
    1. Force and energy—Juvenile literature.   ] 1. Force and energy]   I. Bendick, Karen, 1948—   illus. II. Title.
PZ10.B2952Wk   531'.6   70-187811
ISBN 0-8193-0575-8

CONTENTS

What Makes Things Work?	4
Where Does Energy Begin?	12
Different Kinds of Energy	16
All Energy Changes	24
How Energy Is Used	26
Energy Moves Down a One-Way Street	32
What Happens to Energy?	38
What Do Machines Do?	44
What Is a Machine?	52
Index	63

WHAT MAKES THINGS WORK?

It isn't very hard to find out *how* things work. You can read directions on how to use anything from a pencil sharpener to a computer.

You can take something apart to see how it's made.

You can watch something moving or working, to see what happens.

It's fun to know how things work. But when you know *why* things work, you know a lot more.

What makes a car go?
What pushes a rocket up into space?
What freezes ice cubes?
What heats food when the stove is on?

What puts the lights on when you flip the switch?
What pushes your bike up a hill?

Are there a lot of different answers to those questions?

Or could you answer them all with just one word?

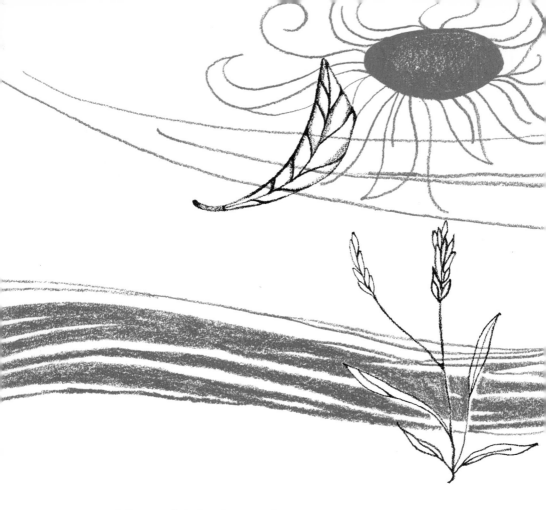

Now, think about *these* questions.
What makes the wind blow?
What makes a plant grow?
What makes a river flow?

Are there a lot of different answers?

Or could you answer all those questions with just one word?

Yes, you could.
The word is

ENERGY

ENERGY
makes a car go.

It makes rivers flow

and green plants grow.

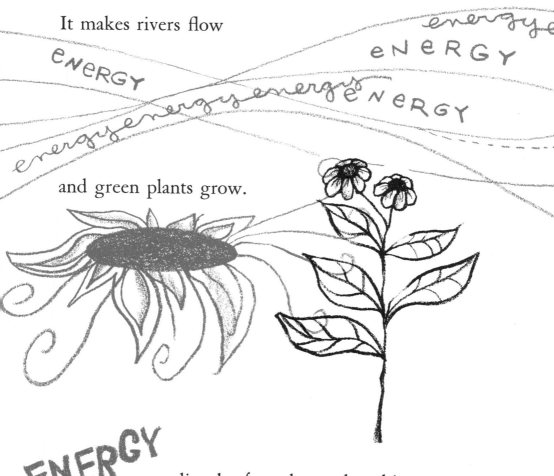

ENERGY supplies the *force* that makes things grow and move and change.

Nothing will grow or change, start or stop, without energy. Energy does all the work in the world.

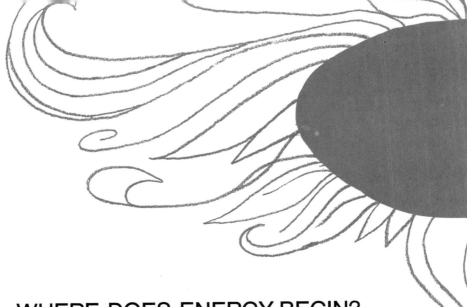

WHERE DOES ENERGY BEGIN?

All the energy on earth begins with the sun. Without the sun, nothing would move or work. Nothing would even be alive.

The sun supplies the energy to make all the green plants on earth grow. When they have sunlight, green plants can make the food they need for living and growing.

Animals can't make food. But animals need food to stay alive. They need it for energy. Some animals eat green plants. Some animals eat the animals that eat green plants.

Without the sun's energy there would be no green plants. And without the plants, no animals.

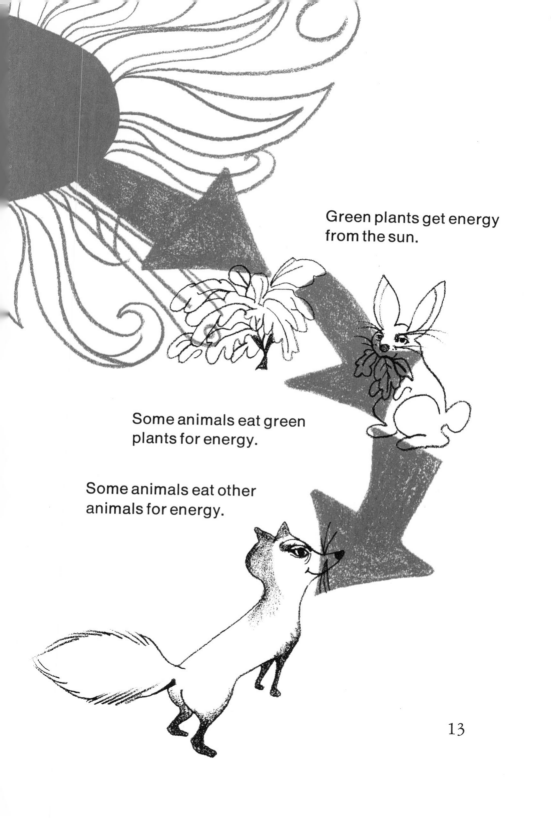

Green plants that lived millions of years ago stored the sun's energy in themselves, when they made food. Even when the plants died and were buried deeper and deeper under the soil, they kept that energy. Slowly, they changed to coal, oil, and gas—the fuels we use to run machines and to make electricity, which runs other machines.

So without the sun, no plants. No fuel to run engines and furnaces, factories and machines. No electricity.

DIFFERENT KINDS OF ENERGY

Light is a kind of energy.
Light from the sun makes plants grow.
Light makes pictures on film.
Light starts and stops some kinds of machines.

and makes a picture on the film at the back of the camera.

Light comes into the camera lens

Heat is a kind of energy.
Heat from the sun makes the weather on earth.
Heat from fuel runs cars and planes.
It cooks food and warms houses.

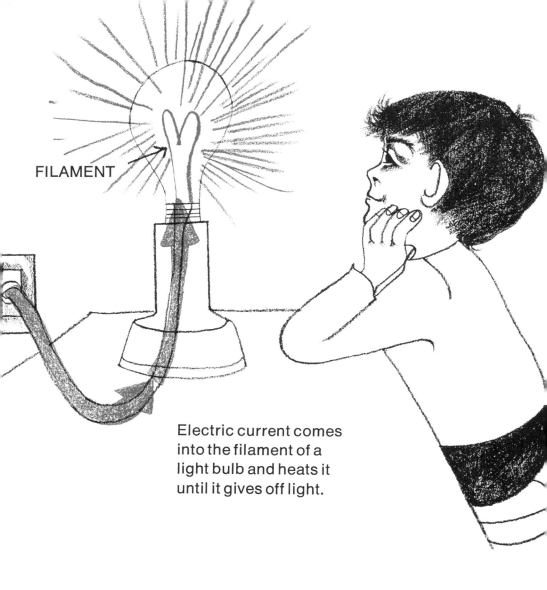

FILAMENT

Electric current comes into the filament of a light bulb and heats it until it gives off light.

Electricity is a kind of energy. Electricity lights lights. It runs computers and motors, television sets and washing machines, refrigerators and toasters.

There is energy in sound.
Sound makes the air move.
That's why people can hear you when you talk—
the moving air jiggles their ears in a special
way.
That's why you hear horns blow and bells ring and
birds sing.

There is energy in your muscles.
That's why you can lift and throw and push and
pull and run and jump.

There is energy in the wind.
It moves clouds and sailboats.
It pushes water into waves.

The energy in water turns wheels and turbines.
Atomic energy makes electricity.
So does magnetic energy.

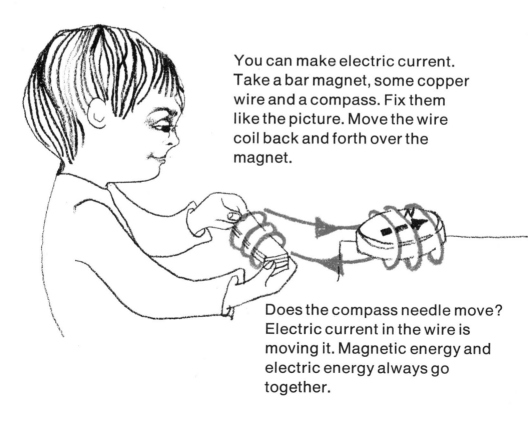

You can make electric current. Take a bar magnet, some copper wire and a compass. Fix them like the picture. Move the wire coil back and forth over the magnet.

Does the compass needle move? Electric current in the wire is moving it. Magnetic energy and electric energy always go together.

And even though all those kinds of energy seem different, one kind can always be changed into another kind.

Heat can be changed into mechanical energy. Burning fuel makes heat that turns water into steam that turns the wheels of an engine that moves a train or a boat.

HOW A STEAM ENGINE WORKS

1. Coal burns in the fire box
2. and makes steam in the boiler
3. which moves the cylinder in the piston
4. and that moves the drive rod, which turns the wheels.

Or the steam might be used to turn a generator. The generator changes mechanical energy into electrical energy.

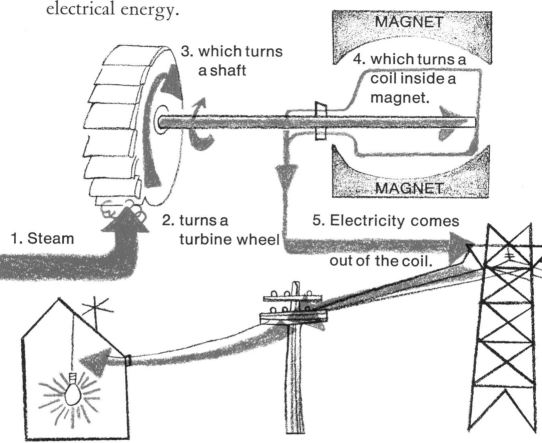

HOW A GENERATOR WORKS

Then the electrical energy might be changed into light in a light bulb, or heat in a toaster, or sound in a radio.

ALL ENERGY CHANGES

There is no kind of energy that can't be changed into another kind. There is no kind of energy that *doesn't* change into another kind, sooner or later.

When energy changes from one kind to another it supplies the force that makes things work.

Think about an apple tree. An apple tree uses energy from the sun to make the food it needs to give it energy for living and growing.

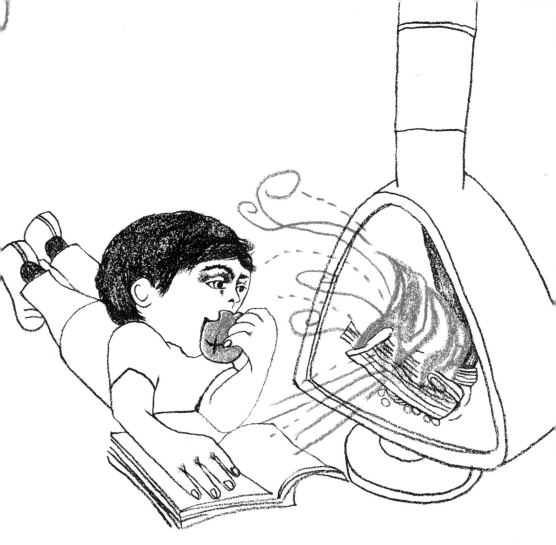

If you eat the apples from that tree your body changes food energy from the tree into muscle energy for you.
If you burn dead branches from the tree in a fireplace, they become heat energy
and light energy.

HOW ENERGY IS USED

Energy doesn't have to be used as it is.
It can be changed into a handier form.
Oil can be used to heat a house, or changed into gasoline to run a car.
Coal and water can be changed into steam.

Energy doesn't have to be used where it is.
It can be moved from one place to another.
Pipelines carry oil.
Trains carry coal.
Wires carry electric current.

Energy doesn't have to be used at any special time. It can be stored for a little time or a long time.

The energy in coal can be stored underground for millions of years.

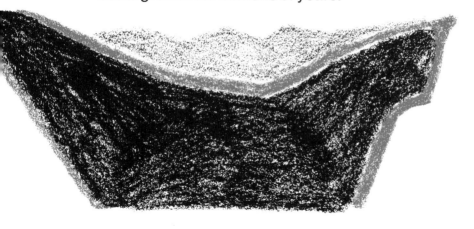

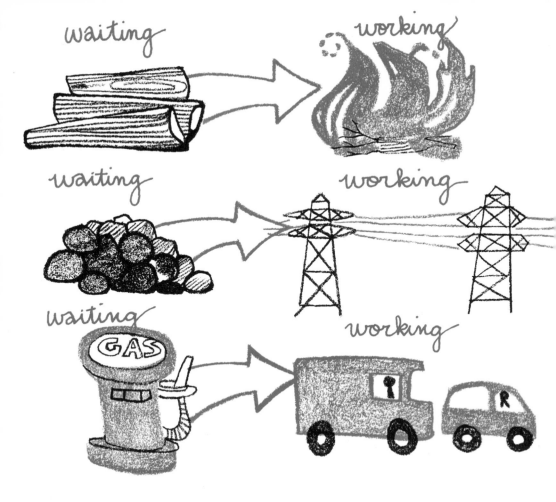

Energy can be stored and waiting, or moving and working.

Energy stored in wood and coal and gas is waiting. When the wood burns and heats a room, or the coal burns and makes electricity, or the gas burns and runs a car, the energy is working.

HOW WATER POWER IS CHANGED TO ELECTRICITY

1. Water is stored behind a dam.

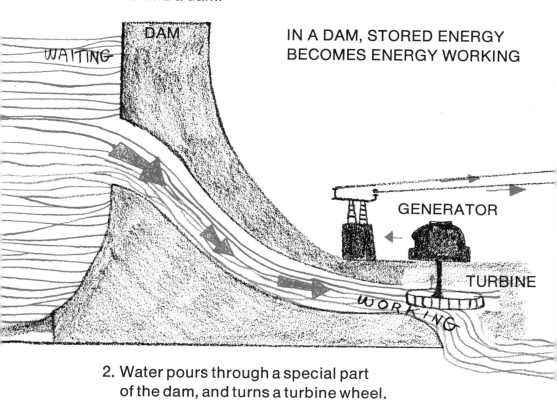

IN A DAM, STORED ENERGY BECOMES ENERGY WORKING

2. Water pours through a special part of the dam, and turns a turbine wheel.

3. The spinning turbine makes electric current in the generator coil.

4. The current is made stronger and put into power lines.

Water behind a dam is stored energy.

Water turning turbine wheels is working.

If you wind up the spring of a toy car, there is energy in the spring, waiting.

When the spring unwinds, the energy is working. The car goes.

If you are hammering a nail,
when the hammer is raised,
the energy in the hammer is stored and waiting.

When you bring the hammer down on the nail,
the energy is working.

ENERGY MOVES DOWN A ONE-WAY STREET

One of the most interesting things about energy is that it always moves down a one-way street.

Energy *always* flows from where there's a lot to where there's less.

You can depend on it.

It happens on the stove. Does heat go from the stove to the pan? Or from the pan to the stove?

Heat has a lot of energy. There is much less energy in cold things.

It happens on your bike. Does energy go from your leg to the pedal? Or from the pedal to your leg?

It happens in a light bulb. Does energy go from the wire into the bulb? Or from the bulb into the wire?

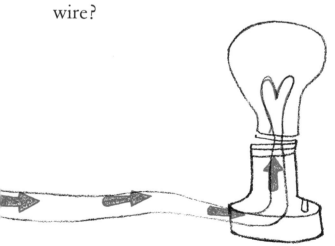

In an engine, heat moves from where the fuel is burning to the parts of the engine that are not hot.

And along the way it supplies the force that pushes things, turns gears, opens valves, and makes the engine run.

In any electrical appliance, when you flip the switch that turns the current on, electricity flows through wires, or vacuum tubes, or transistors.

It moves from where there is a lot of current coming in, to where there is less current, or none.

And when current flows, electricity does all kinds of work.

Energy from the sun travels out through space, away from the sun. It never comes back into the sun.

Light travels out, away from a light bulb.

Your voice travels out, away from you.

Radio waves travel out from a radio station.

Whenever there is a lot of energy in one place, it moves to where there is less. It doesn't matter if the energy is in the form of heat or light, sound or electricity, or anything else.

Every kind of machine we have, whether it works by electricity, gas, muscles, or any other kind of energy, works because energy *always* goes from where there is a lot, to where there is less.

That's what makes things work.

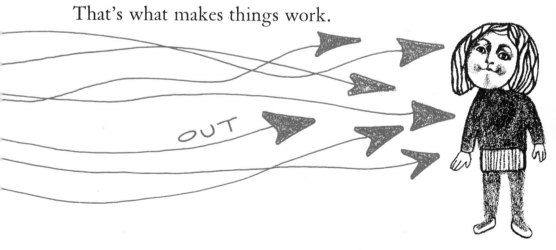

WHAT HAPPENS TO ENERGY?

The sun supplies the energy that makes trees grow.

The energy in a tree can stay there for a year, or a million years.
The tree might be burned as wood, when branches die, or as coal or oil, if it's been underground long enough to change.

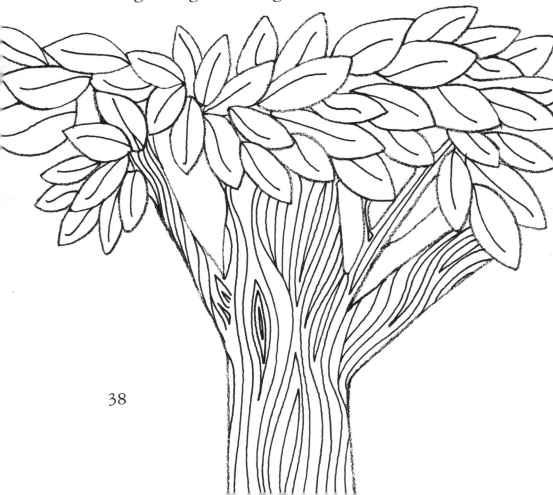

But when the tree is gone, and the heat is gone,
and the light is gone, the energy isn't gone.
It is still around, somewhere.

When energy changes from one form to another, the amount of energy doesn't change.
Energy doesn't disappear.

All the energy there ever was is still around somewhere, but you can't always get at it.

You can change electric current into light, in a light bulb.

But can you catch the light that comes out of the light bulb and use it again?

Some of the light changes into heat. If you hold your hand near the bulb you can feel the heat.

But can you catch that heat to use it again?

Each time energy changes its form, it gets harder to use it again.

A lot of energy from coal

is changed to less electrical energy

which is changed to still less light and heat energy.

Why would anyone try to catch energy to use again? Why not just make more? Don't we have all kinds of machines that make energy?

Machines can't make energy.

Generators and atomic reactors don't make electric power. They change other kinds of energy into electric power.

Electric batteries don't make power. They change chemical energy into electrical energy.

HOW A FLASHLIGHT WORKS

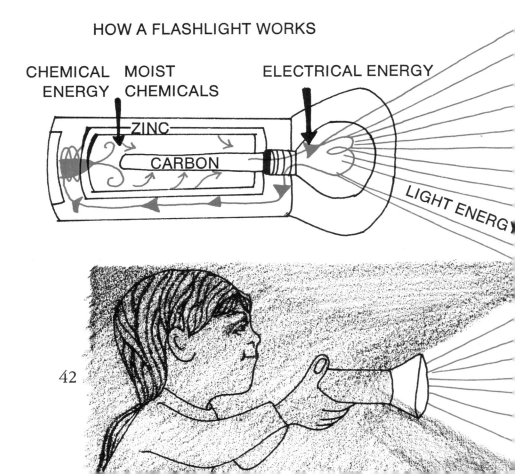

Many space satellites have solar batteries. Solar batteries change light energy from the sun into electrical energy. Electricity runs the satellite's instruments.

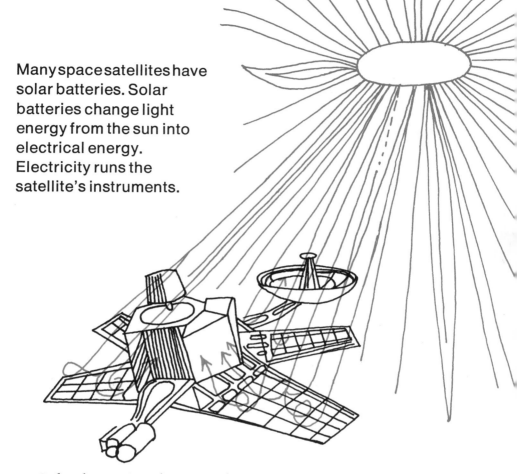

Solar batteries don't make power. They change light energy from the sun into electrical energy.

Steam engines don't make steam power. They change other kinds of energy into steam.

Jet engines and rocket engines don't make power.

Machines can change one kind of energy into another, but they can't make new energy.

WHAT DO MACHINES DO?

Machines do work.
If you look around you can watch machines moving things and people, heating food and houses, and cooling them off, lighting lights and showing pictures, cleaning and sewing,

planting, weeding, and harvesting,
changing one kind of energy into another.
   But they can't *make* energy.

To get any kind of work out of a machine, we have to put energy into it.

A machine can't work without energy.

A car won't run without fuel.

A computer can't figure without electricity.

Even a simple machine like a baseball bat won't work unless someone puts muscle energy into it.

A machine needs energy to change one kind to another.

A generator can't make electricity without fuel.

A furnace can't make heat without fuel.

You put more energy into a spring when you wind it up than the spring puts out as it unwinds.

And we always have to put more energy into a machine than we get out of it.

Every machine wastes energy.
It uses only part of the energy that is put into it.
The rest goes off as heat or noise, smoke or light, or other wasteful things that don't help the work the machine is doing.

All the energy that went into the machine is still around, but we can't get it back to use again. And we can't make new energy.

WHAT IS A MACHINE?

If machines take in energy and use it to do work, are green plants a kind of machine? Maybe you can say that.

Plants take in energy from the sun.
They use that energy to make food for themselves and for all the other living things on earth.
But they always use more energy than there is in the food they make.

Plants take in one kind of energy and change it to another.
But they don't make new energy.

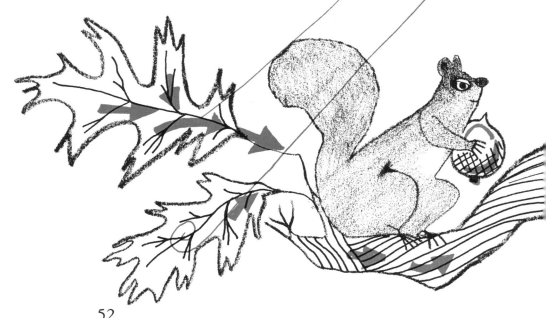

Sunlight falls on green leaves and gives them energy to do their work.

A plant's green leaves make all the food the plant needs for growing, for making fruit and seeds, even for making bark.

Green plants make food that animals use for energy.

Are people a kind of machine?
Maybe you can say that.
People do all kinds of work.
They push and pull and talk and think.
They build things and move things.
They write numbers and make pictures.

But people can't work without taking in energy. All the energy you need for working and growing and even staying alive, comes from the food you eat. But no matter how much work you do, or how much you grow, you always use more energy than you put out.

Some energy is wasted.

And people can't make energy.

Is earth a machine?
Maybe you can say that.
Energy from the sun makes all the parts
of the earth machine work.
Energy from the sun heats the earth.
It grows plants and moves air and water.

There's a lot of work being done and a lot of energy being changed from one form to another by the earth machine.

But only a part of the sun's energy is used to do all those things.

The rest is wasted.

And earth can't make energy.
It can store it, and use it, and change it.
But it can't make it.

What about the sun? Is it a machine?
Maybe you can say that.
The sun makes all the energy it sends through space. But it doesn't make energy out of nothing.
The sun makes energy in itself, out of itself.
It changes part of itself into the energy it sends out into space.

But in billions of years, when the sun's energy is all used up, the sun machine will stop running.
Even the sun can't make energy out of nothing.

Now you know why things work.
Energy makes them work.
They work because energy changes from one form to another. And when it changes it supplies the force that makes things move.

Things work because energy can be stored and used at another time, even in another place.

Things work because energy always moves from where there is a lot to where there is less. And while it is moving we can harness it to do all kinds of jobs.

But you know something else, too.

Once energy is used, we can't get it back.
People use up a lot of earth's energy.
They use coal and oil, gas and water, wood and growing plants.

People waste a lot of the earth's energy.
Do you think we should try to save some?

INDEX

animals, and food, and energy, 12, 13, 53
apple tree, and energy, 24-25
atomic energy, 21
atomic reactors, 42

bicycle, 7, 33

camera, and light, 16
carbon, 42
cars, and energy, 10, 17, 28, 30
chemical energy, 42
chemicals, moist, 42
coal, 27, 41;
    from plants, 15, 38;
    into steam, 26, 27
cold, and energy, 32
compass, 21

dam, 29
drive rod, 22

earth, and energy, 56-57, 62
electric batteries, 42
electric current, 27, 29, 33, 40;
    and light bulb, 18;
    how to make, 21
electrical energy, 21, 23, 41-43
electricity, 18, 23, 28, 29, 35;
    and sun, 15;
    and energy, 21, 28;
    work done by, 35;
    and satellites, 43
energy, 10;
    force supplied by, 10, 24, 60;
    and sun, 12, 13, 14, 36, 38, 43, 52, 53, 56-59;
    different kinds of, 16-23;
    changing of, 22-26, 41-43, 52, 57, 60;
    uses of, 26-31, 52, 54, 55, 57, 61, 62;
    storage of, 27-31, 57, 61;
    moving of, 27, 32-37, 61;
    waiting of, 28-31;
    working of, 28-31;
    catching of, 40-42;
    reuse of, 40-42;
    making new, 43, 45, 51, 52, 55, 57;
    wasting of, 50, 52, 55, 62;
    see also different kinds of
engines, and sun, 15

factories, and sun, 15
filament, 18
fire box, 22
flashlight, how it works, 42
food, 6, 17, 25;
    for plants and animals, 12, 52, 53;
    energy from, 55
fuel, 15, 22;
    see also kinds of
furnaces, and sun, 15

gas, from plants, 15;
    energy in, 28
gasoline, from oil, 26
generators, and energy, 23, 29, 42

heat, 6, 17, 26, 28, 32, 39, 41;
　changing of, 22;
　moving of, 34
heat energy, 23, 25, 32, 41
how things work, 4, 5, 35-37

ice cubes, freezing of, 6

jet engines, 43

light, 7, 16, 39, 40, 41;
　and camera, 16
light bulbs, 18, 23, 33, 37, 40, 41
light energy, 23, 25, 41, 42, 43

machines, and sun, 15;
　and energy, 42, 43, 45-51;
　work of, 44-46;
　kinds of, 52-59
magnet, 21, 23
magnetic energy, 21
mechanical energy, 22, 23
muscle energy, 20, 25, 33

oil, 27;
　from plants, 15, 38;
　into gasoline, 26

people, and energy, 54-55
pipelines, and oil, 27
piston and cylinder, 22
plants, growing of, 8, 12, 53, 56;

　and energy and sunlight, 10,
　　12, 13, 15, 52, 53;
　changing of, 15
power, making of, 42, 43
power lines, 29

river, 8, 10
rocket, 6

solar batteries, 43
sound, and energy, 19, 23
space satellites, 43
springs, and energy, 30, 49
steam, 22, 23, 26, 43
steam boiler, 22
steam engines, 22, 43
sun, and energy, *see under*
　energy

trains, and coal, 27
turbine wheel, 23, 29

water, 21, 26, 29;
　moving of, 56
water power, 29
wheels, turning of, 22, 23, 29
why things work, 5, 60
wind, blowing of, 8;
　energy in, 21
wires, and electric current, 21, 27

zinc, 42

64